Anne-Karoline Distel

100 Birds of Ireland in Watercolour

100birdsofireland@gmail.com

For the Magpie

Preface

This collection of watercolours evolved out of a challenge to myself to paint an Irish bird every day from December 1st to 25th 2017. I had only become interested in birdwatching shortly before I started the challenge, so the variety was quite fascinating to me – especially with the water birds, being a city person.
A friend suggested that it should be made into a book. So I extended the challenge into the new year and ended up with a bit over 130 birds in total. What appears in the book is a selection of 100 of those Irish bird paintings, the ones I like best or considered important in such a collection.
The book is not entirely systematic but begins with the common garden birds - those one can see at the bird feeder; it continues with birds to be found in town and country, and ends with water birds.
Discovering the Irish names and comparing them to their counterparts in the other languages was a particular pleasure – I hope I found the correct names for all of them. There is no index, because the book is meant to be explored rather than being used as a guide.
I hope you enjoy this book as much as I enjoyed painting the birds and learning their names!

Kilkenny, July 2018

Parus major

Great tit ◆ Meantán mór ◆ Kohlmeise

Parus ater

Coal tit ◆ Meantán dubh ◆ Tannenmeise

Carduelis carduelis

Goldfinch ◆ Lasair choille ◆ Stieglitz/ Distelfink

Fringilla coelebs

Chaffinch ◆ Rí Rua ◆ Buchfink

Pyrrhula pyrrhula

Bullfinch ◆ Corcrán coille ◆ Dompfaff/ Gimpel

Carduelis spinus

Siskin ◆ Siscín ◆ Erlenzeisig

Carduelis chloris

Greenfinch ◆ Glasán darach ◆ Grünfink

Passer domesticus

House sparrow ◆ Gealbhan binne ◆ Haussperling

Prunella modularis

Dunnock ◆ Donnóg ◆ Heckenbraunelle

Turdus merula

Blackbird ◆ Lon dubh ◆ Amsel

Turdus pilaris

Fieldfare ◆ Sacán ◆ Wacholderdrossel

Turdus viscivorus

Mistle thrush ◆ Liatráisc ◆ Misteldrossel

Turdus iliacus

Redwing ◆ Deargán sneachta ◆ Rotdrossel

Turdus philomelos

Song thrush ◆ Smólach ceoil ◆ Singdrossel

Apus apus

Swift ◆ Gabhlán gaoithe ◆ Mauersegler

Regulus regulus

Goldcrest ◆ Cíorbhuí ◆ Wintergoldhähnchen

Aegithalos caudatus

Long-tailed tit ◆ Meantán earrfhada ◆ Schwanzmeise

Riparia riparia

Sand martin ◆ Gabhlán gainimh ◆ Uferschwalbe

Delichon urbica

House martin ◆ Gabhlán binne ◆ Mehlschwalbe

Plectrophenax nivalis

Snow bunting ◆ Gealóg shneachta ◆ Schneeammer

Emberiza schoeniclus

Reed bunting ◆ Gealóg ghiolcaí ◆ Rohrammer

Phylloscopus collybita

Chiffchaff ◆ Tiuf-teaf ◆ Zilpzalp

Phylloscopus throchilus

Willow warbler ◆ Ceolaire sailí ◆ Fitislaubsänger

Acrocephalus schoenobaenus

Sedge warbler ◆ Ceolare cíbe ◆ Schilfrohrsänger

Locustella naevia

Grasshopper warbler ◆ Ceolaire casarnaí ◆ Feldschwirl

Fringilla montifringilla

Brambling ◆ Breacán ◆ Bergfink

Loxia curvirostra

Red crossbill ◆ Crosghob ◆ Fichtenkreuzschnabel

Anthus pratensis

Meadow pipit ◆ Riabhóg mhóna ◆ Wiesenpieper

Carduelis cannabina

Linnet ◆ Gleoiseach ◆ Bluthänfling

Sylvia communis

Whitethroat ◆ Gilphíb ◆ Dorngrasmücke

Sylvia atricapilla

Blackcap ◆ Caipín dubh ◆ Mönchsgrasmücke

Oenanthe oenanthe

Wheatear ◆ Clochrán ◆ Steinschmätzer

Bombycilla garrulus

Waxwing ◆ Síodeiteach ◆ Seidenschwanz

Saxicola rubetra

Whinchat ◆ Caislín aitinn ◆ Braunkehlchen

Saxicola rubicola

Stonechat ◆ Caislín cloch ◆ Schwarzkehlchen

Phoenicurus phoenicurus

Redstart ◆ Earrdheargán ◆ Gartenrotschwanz

Carduelis flammea

Redpoll ◆ Deargéadan ◆ Birkenzeisig

Sturnus vulgaris

Starling ◆ Druid ◆ Star

Certhia familiaris

Treecreeper ◆ Snag ◆ Waldbaumläufer

Alauda arvensis

Skylark ◆ Fuiseog ◆ Feldlerche

Ramphastos toco

Toco toucan ◆ Tucán ◆ Riesentukan

Ficedula hypoleuca

Pied flycatcher ◆ Cuilsealgaire alabhreac ◆
Trauerschnäpper

Muscicapa striata

Spotted flycatcher ◆ Cuilsealgaire liath ◆ Grauschnäpper

Pica pica

Magpie ◆ Snag breac ◆ Elster

Passer montanus

Tree sparrow ◆ Gealbhan crainn ◆ Feldsperling

Columba palumbus

Wood pigeon ◆ Colm coille ◆ Ringeltaube

Columba livia

Rock pigeon ◆ Colm aille ◆ Felsentaube

Motacilla cinerea

Grey wagtail ◆ Glasóg liath ◆ Bergstelze

Motacilla alba

Pied wagtail ◆ Glasóg shráide ◆ Bachstelze

Corvus corone cornix

Hooded crow ◆ Caróg liath ◆ Nebelkrähe

Corvus monedula

Jackdaw ◆ Cág ◆ Dohle

Corvus corax

Raven ◆ Fiach dubh ◆ Kolkrabe

Tyto alba

Barn owl ◆ Scréachóg reilige ◆ Schleiereule

Falco columbarius

Merlin ◆ Meirliún ◆ Zwergfalke

Buteo buteo

Buzzard ◆ Clamhán ◆ Bussard

Lagopus lagopus

Willow ptarmigan ◆ Tarmachan ◆ Moorschneehuhn

Crex crex

Corn crake ◆ Traonach ◆ Wachtelkönig

Phasianus colchicus

Pheasant ◆ Piasún ◆ Fasan

Perdix perdix

Grey partridge ◆ Patraisc ◆ Rebhuhn

Coturnix coturnix

Quail ◆ Gearg ◆ Wachtel

Tachybabtus ruficollis

Little grebe ◆ Spágaire tonn ◆ Zwergtaucher

Aythya ferina

Pochard ◆ Póiseard ◆ Tafelente

Anas acuta

Pintail ◆ Biorearrach ◆ Spießente

Aythya fuligula

Tufted duck ◆ Lacha bhadánach ◆ Reiherente

Anas penelope

Wigeon ◆ Lacha Rua ◆ Pfeifente

Bucephala clangula

Goldeneye ◆ Órshúileach ◆ Schellente

Anas strepera

Gadwall ◆ Gadual ◆ Schnatterente

Fulica atra

Eurasian coot ◆ Cearc cheannann ◆ Blesshuhn

Tadorna tadorna

Shelduck ◆ Seil-lacha ◆ Brandgans

Cygnus cygnus

Whooper swan ◆ Eala ghlórach ◆ Singschwan

Cygnus olor

Mute swan ◆ Eala balbh ◆ Höckerschwan

Branta leucopsis

Barnacle goose ◆ Gé ghiúrainn ◆ Nonnengans

Anser albifrons

White-fronted goose ◆ Gé bhánéadanach ◆ Blessgans

Branta canadensis

Canada goose ◆ Gé cheanadach ◆ Kanadagans

Branta bernicla

Brent goose ◆ Gé cadhan ◆ Ringelgans

Anser anser

Greylag goose ◆ Gé ghlas ◆ Graugans

Egretta garzetta

Little egret ◆ Éigrit bheag ◆ Seidenreiher

Alcedo atthis

Kingfisher ◆ Cruidín ◆ Eisvogel

Cepphus grylle

Black guillemot ◆ Forache dhubh ◆ Gryllteiste

Fratercula arctica

Puffin ◆ Puifín ◆ Papageitaucher

Larus argentatus

Herring gull ◆ Faoileán scadán ◆ Silbermöwe

Morus bassanus

Gannet ◆ Gainéad ◆ Basstölpel

Sterna hirundo

Common tern ◆ Geabhróg ◆ Flussseeschwalbe

Pluvialis squatarola

Grey plover ◆ Feadóg ghlas ◆ Kiebitzregenpfeifer

Pluvialis apricaria

Golden plover ◆ Feadóg bhuí Áiseach ◆ Goldregenpfeifer

Charadrius hiaticula

Ringed plover ◆ Feadóg chladaigh ◆ Sandregenpfeifer

Vanellus vanellus

Northern lapwing ◆ Pilibín ◆ Kiebitz

Tringa totanus

Common redshank ◆ Cosdeargán ◆ Rotschenkel

Tringa nebularia

Common greenshank ◆ Laidhrín glas ◆ Grünschenkel

Gallinago gallinago

Snipe ◆ Naoscach ◆ Bekassine

Rallus aquaticus

Water rail ◆ Ralóg uisce ◆ Wasserralle

Numenius arquata

Curlew ◆ Crotach ◆ Großer Brachvogel

Numenius phaeopus

Whimbrel ◆ Crotach eanaigh ◆ Regenbrachvogel

Limosa limosa

Black-tailed godwit ◆ Guilbneach earrdhubh ◆ Uferschnepfe

Philomachus pugnax

Ruff ◆ Rufachán ◆ Kampfläufer

Arenaria interpres

Turnstone ◆ Piardálai trá ◆ Steinwälzer

Calidris alpina

Dunlin ◆ Breacóg ◆ Alpenstrandläufer

Calidris canutus

Red knot ◆ Cnota ◆ Knutt

Calidris alba

Sanderling ◆ Luathrán ◆ Sanderling

Printed in July 2019
by Rotomail Italia S.p.A., Vignate (MI) - Italy